THINKING
THROUGH
THE
ENERGY
PROBLEM

THINKING THROUGH THE ENERGY PROBLEM

Thomas C. Schelling

working with the
DESIGN COMMITTEE ON LONG-RANGE ENERGY POLICY
comprised of
HENRY B. SCHACHT, *Chairman*
A. ROBERT ABBOUD
ROBERT C. HOLLAND
FRANKLIN A. LINDSAY
WILLIAM F. MAY
JOHN C. SAWHILL

Committee for Economic Development

Library of Congress Cataloging in Publication Data

Schelling, Thomas C.
    Thinking through the energy problem.

    (A Supplementary paper of the Committee for Economic Development; 42)
    1. Energy policy—United States. 2. Petroleum products—Prices. I. Committee for Economic Development. Design Committee on Long-Range Energy Policy. II. Title.
HD9502.U52S32   333.7   79-4583
ISBN 0-87186-242-5

First Printing: March 1979
Printed in the United States of America
Design: Stead, Young & Rowe, Inc.
Price: $5.00

COMMITTEE FOR ECONOMIC DEVELOPMENT
477 Madison Avenue, New York, N.Y. 10022
1700 K Street, N. W., Washington, D.C. 20006

CONTENTS

FOREWORD

The strength and stability of the U.S. economy depend very heavily on an effective solution to the nation's energy problem. *Thinking Through the Energy Problem* offers a fresh approach to the multifaceted issues surrounding energy and a solid framework within which energy policies, present and future, can be judged.

The close association between energy and the economy has long been recognized by the Committee for Economic Development. In 1973, months before the Arab oil embargo, a CED Subcommittee was at work formulating proposals both for stimulating energy production and curbing energy demand. CED's findings and recommendations appeared in the policy statement *Achieving Energy Independence,* published in 1974.

Since that initial report, CED's Research and Policy Committee has published three policy statements that respond to various aspects of the nation's energy problem: *International Economic Consequences of High-Priced Energy* (1975), *Nuclear Energy and National Security* (1976), and *Key Elements of a National Energy Strategy* (1977). The 1977 statement warned that the intensifying public debate over energy was in danger of becoming "so enmeshed in details that fundamental domestic and international considerations may become obscured."

With this warning in mind, CED commissioned Professor Thomas C. Schelling of Harvard University to prepare a study designed to help public officials and private citizens think through the energy problem in a rational and objective manner and to identify certain fundamental principles. It was clear from the outset that the purpose of such a study would be to construct a conceptual framework for evaluating energy policy, not to devise a set of specific recommendations.

Working closely with Professor Schelling was the Design Committee on Long-Range Energy Policy, a small task force of CED trustees whose names and affiliations appear on page v. The Design Committee, assisted by experts from business and academe, met with Professor Schelling monthly over a period of more than a year, commenting on the successive stages of his analysis and exchanging ideas as to its applications. *Thinking Through the Energy Problem* is a distillation and refinement of concepts that emerged from this process.

Professor Schelling focuses on oil imports as the principal connection between United States domestic energy policy and a multitude of energy-related strategic and foreign policy issues. The question of oil imports, he maintains, deserves special attention in the design of energy policy. Indeed, how oil imports are treated becomes a key element in the formulation of American energy strategy.

Instead of proposing specific solutions, the study suggests new ways of looking at the connections among energy issues and then explores the implications of those connections in the choice of policy mechanisms.

Among the myriad issues surrounding energy, Professor Schelling identifies one that is paramount: *price.* The prices that people are willing to pay for existing fuels, he states, help determine how much new fuel can be developed and at what cost. "Keeping fuel prices artificially below the replacement cost of the fuels being used," he argues, "subsidizes excessive consumption, inhibits exploration and development of supply, and misrepresents the worth of technological changes that economize energy." Price regulation, he maintains, may disguise the ways the costs of fuel are paid and who pays them, but it does not reduce those costs.

Professor Schelling finds that the energy-related costs of environmental protection are an important factor in the rise of energy costs; yet the price system does not reflect these costs. Moreover, these costs will grow substantially, he believes, unless environmental protection is treated as an "economic choice," with costs related to benefits, rather than as a "technological absolute."

Recognizing that the market system cannot respond to all environmental and foreign policy concerns, Professor Schelling nevertheless concludes that the market's virtues of flexibility and adaptability are our best resources in dealing with most of the risks and uncertainties in the energy picture.

Thinking Through the Energy Problem is published as a CED Supplementary Paper, and as is traditional for such papers, the author takes responsibility for its contents. In this case, however, the study was discussed, debated, and its contents unanimously endorsed by the trustees who comprised the Design Committee on Long-Range Energy Policy. Its publication was strongly endorsed not only by this group, but also by CED's sixty-member Research and Policy Committee in this language:

> This analysis of the nature of the energy problem is being made available by CED as a framework for addressing energy policy. It does not contain specific recommendations. It differs in that respect from CED policy statements, which do contain recommendations that have been voted on by CED's Research and Policy Committee and that also may contain dissents and reservations. Publication of this statement, prepared by Professor Thomas C. Schelling working with a small Committee of CED trustees, has been endorsed by the Research and Policy Committee as a fundamental and constructive perspective for its ongoing consideration of energy policy and is published so that it may serve that purpose for others as well.

Robert C. Holland
President
Committee for Economic Development

1

THE NATURE
OF THE
ENERGY PROBLEM

Most of our energy, other than sunlight, comes from burning the fossil fuels—oil, gas, and coal. These fuels are burned in homes to keep warm, in engines to drive machinery, in electricity plants to make steam, and in mills and factories to make paper, glass, steel and fertilizer. Some of our electric energy comes from nuclear reactors and from rainwater flowing downhill. Small amounts of energy are in the food we eat, and huge amounts radiate from the sun to keep the planet warm, make the vegetation grow, and produce the rain and wind and the light we see by in the daytime.

The quantities are impressive. If the heat value of coal and gas is converted to oil equivalents, every man, woman and child in America consumes on the average, directly and indirectly, more than his body weight in petroleum every two days. A fifth of this fuel is imported oil, costing about $40 billion a year. The value of all the fuel is about one-twentieth the value of the gross national product (GNP).

As technology changes, so does the use of energy. As fuel prices go up or down, fuel is used sparingly or liberally. The trend is for energy use, measured in heat content, to rise with both population and per capita income. With the GNP doubling about every twenty years, the use of energy might double in about twenty-five years and quadruple in half a century, if energy prices stay even with other prices.

There are two inescapable facts about the supplies of fossil fuels. One is that the quantities were determined millions of years ago in the geological history of the planet; there will never be any more than there already was by the time some primitive person discovered that a lump of coal would burn. The other is that they are strikingly unequal in their geographical distribution: half the known petroleum reserves but only one-hundredth of the world's population are in the countries— until recently exceedingly poor, now strangely rich—that border the Persian Gulf.

With the demand for fuels forever growing, potentially quadrupling in half a century, and with supplies that were fixed before history began, it is tempting to ask how long they will last. More straightforwardly, how much oil, gas and coal is *known* to exist, how much can be *economically* extracted, and how do those quantities compare with *projected* consumption? Even the answers to these questions are speculative. The resources are underground in complex formations; and, especially with oil, during most periods, more was discovered than extracted, so that the known amount remaining continually increased. But so did consumption. Consequently, there was always a decade or two of proven reserves awaiting extraction. The urgency of exploration is inversely related to reserves; with a lead time of ten years necessary to find and exploit new supplies, search and discovery are more highly motivated if known reserves will last only ten years than if they will last forty.

As the earth is more explored it becomes less likely that new large deposits will be found. With the accumulation of experience, estimates of likely deposits, even where no drilling has taken place, should become more exact. But actually there remain huge uncertainties about undiscovered oil and gas, even coal. Parts of the earth's surface, especially under the oceans, have not been explored. Exploration of some areas is beyond present technology. There are depths to which drilling has not gone even in familiar areas. Estimates of the oil and gas that may be found at depths and in locations yet unexplored are only guesses. Some of the guesses are highly encouraging and cannot be refuted with confidence; some are discouraging.

We have not closed the frontier on the discovery of new fossil fuel reserves. Indeed, significant new discoveries are continually being announced. But recent trends and general reasoning strongly suggest—not prove, just strongly suggest—an end to the era when the exponential growth of demand would be continually matched by commensurate growth in the discovery of new easily accessible reserves, and a decline in the ratio of known supply to effective demand *at today's prices.*

THE PRICE DIMENSION

There is a fundamental reason why the question of how much petroleum, gas, and coal exists does not admit a definite answer. It is that *the amounts of fuel that can be economically exploited depend on the prices people will pay for them.* Even from abandoned deposits there is oil to be had at higher extraction costs. Deeper wells can be drilled; oil can be obtained from the ocean bed; it can be brought expensively by pipeline across the entire state of Alaska. Natural gas in huge quantities may be available several miles under the surface. Eventually shale and tar sands can yield large quantities of fuel but at costs that have not been, and still are not, competitive with the common fuels.

Technologies of extraction that are still undeveloped or unproven will become available to bring additional supplies of gas and liquid fuel onto the market in years or decades to come. These technologies will be more urgently developed, the higher the prospective prices of oil and gas.

Coal illustrates the same principle. In the United States alone the amount of potentially combustible coal is estimated at more than a trillion tons, enough to last a thousand years at today's rate of extraction. But successive billions or tens of billions of tons will be progressively more expensive because of quality, depth and thickness, location, and, especially, the environmental effects of mining, transporting, and burning it.

The likely consequence is that fuel costs will not decline

during the coming decades but will rise as the growth of demand confronts the depletion of the most accessible, and least environmentally damaging, supplies of oil, gas and coal. This prospect is often described as the overtaking of supply by demand and the development of a "gap". But that image of the "energy problem" neglects what is central to the relation between supply and demand: the *costs* at which fuels can be produced, and the *values* of the fuels in their myriad uses. More and more expensive sources of supply will have to be used; rising demand will provide the market for them, at higher prices to cover their higher costs. The "energy problem" is not best described as a comparison of demand with supply and the emergence of a gap between them, but as a prospective rise in the cost of fuel. Only if prices are continually regulated below the cost of producing the fuel will an actual "gap" be observed.

That is the medium-term energy problem, during our and our children's lifetimes. It is not that the world will run out of fuel on some uncertain date, or that demand will outstrip supply and create a finite shortage. It is that fuel has become more expensive and is likely to become increasingly so. The energy problem is not to keep the price of fuel from rising. It is to meet the rising economic cost of fuel with policies that minimize the burdens, allocate them equitably, avoid disruptions in the economy, and keep the costs from rising more than necessary.

CHARACTERISTICS OF THE ENERGY PROBLEM

There are five characteristics of this problem to be highlighted. One has been mentioned—the enormous *uncertainty* about the quantities of fuel that will become available at different costs, and with new technologies, in years to come. There is some probability that new discoveries, or new technologies for extraction, will dramatically enhance supply within the next decade or two, postponing or attenuating the rise in the cost of fuels. It is extraordinarily difficult to devise policies in the face of the good news that there is one chance in ten that we shall discover unexpected great wealth within the decade.

A second characteristic is the *long lead time* necessary for almost any development or adaptation to price changes. New supplies of oil, gas, and coal take a decade or more to develop. Conservation of energy often requires replacing plant and equipment, whether for generation of electricity or for trucks, aircraft and automobiles. New technologies, like liquefaction or gasification of coal, involve not a decade but two in their development and commercialization. For purposes of policy we are already in the 1990s. Today's energy decisions will mainly affect supply, conservation, or new technology ten or fifteen years hence.

A third characteristic is that extracting fuel, transporting it, and burning it affect health, safety, and land use, the esthetics and productivity of the terrain, the sociology and demography of remote areas. It is unlikely that concern with the *environment*, greatly enhanced over the past ten years, will diminish in the future. The effects on health and productivity, if not the esthetics, once discovered are not likely to be suppressed. Another million coal cars continuously moving through towns and countryside will be more than an esthetic nuisance.

A fourth characteristic of energy is its impact on the *balance of payments*. About a fifth of the fuel we consume in the United States is imported, and our balance of payments and the value of the dollar are affected by whether we import forty billion dollars of foreign oil each year or a hundred billion.

A fifth characteristic is that an even larger part of the energy consumed by Japan and Western Europe is imported from a small number of oil producing countries, mostly in the Middle East. The supply is susceptible to sudden *disruption*, motivated politically or economically, in peacetime or warlike circumstances. For the United States a sudden disruption would be serious but not devastating. For some countries, like Japan, cessation of overseas oil supply could be a disaster.

The United States is blessed with large quantities of oil, gas, and coal, producing 80 percent of the fuel it consumes and undoubtedly is capable, though not on short notice, of meeting its most urgent needs out of domestic supplies. Most countries, developed and underdeveloped, are more vulnerable than the United States to an interruption in oil imports.

WORLD PROBLEMS
RELATED TO ENERGY

There is a complex of world problems related to energy that transcend the geology and the economics. Many but not all of them relate to the concentration of petroleum reserves in the Middle East. The roles of Cuba, Russia, China and the United States in the Somali-Ethiopian conflict on the "Horn of Africa" reflect the strategic significance of oil. The sale of advanced military aircraft to Saudi Arabia is part of the world energy situation. Overflight and landing rights for military supply of Israel were affected during the war of 1973 by the European politics of Middle East oil. Arab-Israeli peace negotiations and Chinese-Japanese trade relations are involved in world energy. The prospects for world trade in plutonium fuel for nuclear reactors were affected by the price of oil in the early 1970s. And the stability of capital markets, even the solvency of banking systems, have appeared to be threatened in some degree by the concentrated financial flows attendant on the Middle East oil trade.

Worldwide problems related to energy are somewhat diffuse, hypothetical, and intertwined with non-energy-related trends and conflicts. But some can be identified as potential crises. Today oil is practically the only economic resource that one can imagine leading to war. Oil has strained America's relations with allied countries. A protracted interruption in delivery of oil to the rest of the world is one of the few genuine economic calamities that come to mind. Control of Middle Eastern oil by the Soviet Union might be construed as such a threat to the viability of Japan and Europe as to be intolerable.

*   *   *

For the far future it may be wise to anticipate sources of energy quite different from the traditional fossil fuels. Sunlight, which can begin to meet some of our energy needs immediately in space and water heating, can eventually produce electricity. Nuclear fusion is a possible source of electricity by the middle of the next century. And electricity itself can produce clean fuel in the form of liquid hydrogen. Some time in the next century it may be necessary, despite the continued availability of high-cost fossil fuels, to abandon primary reliance on them for reasons related to climate, health and environmental damage.

THE POLITICS OF PRICES

A final "energy problem" needs to be anticipated, one not based on geology or economics. It is the politics of energy. There is a widespread tendency to view fuel price increases not as reflections of genuine costs, not as an adaptive response of the market to the need for conservation and enhanced supply, but as the problem itself. Price regulation can disguise the ways the costs of fuel are paid and who pays them. It can redistribute costs but it does not reduce them. Keeping fuel prices artificially below the replacement cost of the fuels being used subsidizes excessive consumption, inhibits exploration and development of supply, and misrepresents the worth of technological changes that economize energy. *If prices are considered the problem, rather than part of the solution, we shall only aggravate problems that are going to be difficult enough.*

2

THE SIZE
OF THE
ENERGY PROBLEM

U.S. energy problems can be divided into two groups, one worldwide and the other domestic. The worldwide problems are multifarious, complex, and rarely limited to energy. But to *clarify their relation* to domestic U.S. energy a common connection deserves emphasis.

To see this connection, consider first some issues and proposals for domestic energy policy, then the worldwide issues that revolve around or interact with energy, and finally the connections between these two domains—the energy channels through which domestic and worldwide issues impinge on or interact with each other.

A Sampling of Domestic Energy Issues

- Price controls on gas and crude oil
- Taxes to induce conversion to coal
- Subsidies to energy-conserving technologies
- Water rights
- Tax benefits for home insulation
- Peakload electricity pricing
- Auto engine design
- Offshore leasing
- Gasoline mileage standards
- Auto emission standards
- Siting of power plants
- Cogeneration of electricity
- Mass transit
- Strip mine regulation
- Coal transport rights-of-way
- Coal mine health and safety standards
- Energy labelling of electric appliances
- Allocation of wellhead tax proceeds
- Right turn on red light
- Disposal of nuclear wastes
- Liquefaction and gasification of coal
- Development of solar-electric technology
- Alternatives to energy-intensive fertilizers

The domestic energy issues listed in the accompanying box range from important to trivial, from technical to social, from supply to demand, from industrial to household, and from federal to state and local, judicial, and voluntary. One way or another they all involve increasing the *availability* of energy or its substitutes, reducing the *need* for energy or the waste of it, mitigating *side effects* of energy production or use, redistributing the public and private *costs* of using energy or doing without it, rearranging *incentives*, or discovering or disseminating pertinent *knowledge*. They reflect the pervasiveness of energy in a modern economy and a network of substitution possibilities that keep all the listed items at least indirectly connected with each other.

A sampling of foreign-policy issues that involve energy directly, or that derive their importance or their difficulty from their connection with energy are described in the box below.

A Sampling of Foreign Policy Issues Related To Energy

- The Soviet role in the Middle East
- Proliferation of nuclear-explosive materials
- Overflight and landing rights for military supply in the Middle East
- Advanced armaments for Persian Gulf countries
- Japanese-Arab relations
- Economic development of Third World countries
- Chinese-Japanese trade relations
- Stability of international capital markets
- Brazilian-American nuclear energy relations
- French cooperation with NATO
- Cuban intervention in the Horn of Africa
- Arab-Israeli peace negotiations
- The danger of a disrupted oil supply as a political move, as an economic strategy, or from sabotage, war, or overthrow of regimes in the Middle East.

With only a little simplification it can be said that these two domains, the domestic energy and the worldwide energy-related, have a single major intersection: U.S. oil imports. Except for that connection, the domestic *energy* issues and the world *energy-related* issues impinge on each other little and only indirectly.

The rest of the world feels our energy policies through the oil we import. It makes little difference to energy prices abroad, to Japanese policy toward the Middle East, to the economic development of India or the nuclear development of Brazil, to the relations between Iran and Saudi Arabia, or to almost any important world energy-related issue, what we do specifically about allocating natural gas, taxing gasoline, insulating new buildings, leasing offshore oil, siting of nuclear reactors, peak-load pricing of electricity, or metering heat in individual apartments—except for what these do to the oil we import. All the consequences get translated into the common currency of British thermal units (BTUs), and their impact is transmitted abroad mainly through the BTUs we import in the form of oil.

The issues in those two domains, the world and the domestic, contrast in their potential gravity. Some of the worldwide problems can be identified as potentially of crisis proportion, carrying at least the possibility of catastrophe. They are undoubtedly what the President of the United States had in mind when in April 1977 he referred to the energy crisis as the greatest challenge, short of prevention of war itself, that we may face in our lifetime. Whether we agree or disagree on the likelihood that world energy problems will lead to some kind of catastrophe, most of us can at least imagine what some of the potential catastrophes might be. It is difficult to reach a common perspective on grave events whose likelihood may be only one chance in five, or one in fifty.

In contrast, the domestic problem of accommodating to the rising cost of energy is not a mortal crisis. It is a serious problem, large but finite; its boundaries can be estimated. It is made more serious by the possibility of sudden disruption in the availability of imported oil, or drastic increases in the purchase price of imported oil; but limits can be estimated on the harm that unexpected disturbances in supply could cause and on the costs of programs to minimize vulnerability.

The prospective continuing rise in the cost of fuel is not so much a problem as a condition. The problem is how to devise short-range policies and long-range strategies for absorbing those price increases into an economy that can grow, in size and in productivity, without inflation and with a reasonable distribution of the benefits of growth among the population.

The rising cost of fuel is one of several serious economic conditions to which we must accommodate during the coming decades. In magnitude it is not altogether unlike the prospect of the medical care industry's consuming not 7 or 8 but 12 or 15 percent of our GNP in another twenty years if we fail to moderate demand and improve supply. Like the aging of the U.S. population and the rising costs of Social Security that will result, the increasing cost of fuels is bad news. The question is: how bad? It is important to get an assessment of its severity.

If it is truly less grave than the global problems—or less grave at its worst than the global problems would be at their worst—it is important to keep the two sets of problems distinct in our minds. We shall exaggerate the domestic difficulties if we approach them with a sense of crisis in the image of Soviet-American confrontation in the Persian Gulf, or the spread of nuclear-weapons material to adventurous governments. And we shall misconstrue the nature of the multifarious worldwide problems if we think of them as primarily endangering our oil supply, or, conversely, think they are to be managed or solved primarily by our energy policies.

The rest of this chapter is an assessment of the likely scope of the domestic energy problem. The role of imported oil—quantity, price, and dependability—is part of that assessment. The special role of oil imports as the main energy supply not under U.S. control, and as the largest and most direct channel of influence U.S. domestic energy has on the rest of the world, will be examined in the chapter that follows.

THE COST IN PRODUCTIVITY

To approximate the likely scope of the domestic problem, the following rough calculation can be made. At twice the present cost of imported *liquid fuel* we can probably have adequate supplies of coal-based liquid fuels, fuels from shale, from old wells by means of enhanced production, and from the oil and gas that may be discovered at depths and distances that will become worthwhile at a market price equivalent to, say, $30 per barrel at 1978 prices. Having these supplies at such prices could take fifteen or twenty years, but existing supplies and new discoveries of conventional gas and oil will be more than adequate, at prices between today's OPEC price and twice that price, to meet our needs during the interim. Nuclear power and coal will provide electricity at prices that will likely rise but will not double, even if the costs of coal and uranium double. Aside from short-run disturbances to the overseas supply of imported oil, then, a reasonable upper boundary on what may be in store for the cost of fuel is another doubling between now and the end of the century.

If prices do not change, the growth in energy use is less than proportionate to the growth of the GNP. The GNP is likely to reach twice its present level just after the year 2000. Without any change in energy prices, the use of energy might then be expected to be 80 or 85 percent greater than it is today.

The effect on demand of a doubling of fuel prices is highly conjectural. Except very recently, fuel prices have not risen substantially in peacetime. The recent experience is limited as evidence because most responses to higher fuel prices take time. Many of the responses depend on expectations of future prices; they involve durable equipment and other long-lived investments, even changes of location. Consumption patterns and production technologies in countries that have had much higher fuel prices than the United States are suggestive but rarely comparable. The effect by the year 2000 would depend on the profile of price increases during the interim years. With a doubling of prices, a conservative guess might be a reduction in energy use by 15 or 20 percent below the level associated with unchanged prices. (A conservative estimate is justified because much of the response may be long delayed.)

So if fuel prices were to double again, in relation to the general price level, by the time the GNP had doubled around or just after the year 2000, the use of energy in all its forms could be projected at about 15 or 20 percent below 1.8 times today's use of energy, or roughly one-half more than today's consumption.

An increase of that magnitude over the next twenty-five years, with an adequate mix of solid, liquid, gas and electricity, should be forthcoming at prices up to twice as high as today. Thus when the required quantities are taken into account, reflecting both the negative response of demand to higher prices and the positive response to higher GNP, a possible doubling of the average cost of fuel over the next couple of decades appears to be a fair upper limit for our calculation of how severe the domestic energy cost increase needs to be.

Raw energy, fuel at mine or wellhead or tanker dock, currently constitutes close to 5 percent of our gross national product. Because oil and gas from older producing wells have regulated prices substantially below their replacement costs, a more realistic estimate of the current cost of fuel in our GNP might be around 6 to 7 percent. With no change in the relative costs of fuels, the corresponding fuel figure for the doubled GNP of 2000-2005 would be about 5 to 6 percent. A doubling of prices with no reduction in demand would thus add 5 to 6 percent of GNP to the cost of fuel. With the 15 to 20 percent reduction in demand, the added cost due to a doubling of prices would be equal to something less than 5 percent of GNP.

That is about the size of the medium-term "energy problem". It may be equivalent to permanently subtracting up to 5 percent from real GNP by about the year 2000. The cost would show up as reduced productivity in the industries producing fuel (and in the terms of trade with oil exporting nations). This much of our GNP would simply disappear into the costs of producing energy and, to some extent, the costs of getting along with less. Most of it would be the higher cost of extracting, delivering, or transforming energy, or abating the environmental damage; some of it would be the extra costs of accommodating, through fuel-saving technologies and consumption patterns, to the higher relative price of fuel.

This, of course, is merely a crude upper-bound estimate that pays no attention to the mix of coal, oil, gas, liquids or gases derived from coal, or nuclear or hydroelectric power. It is a rough estimate of gross magnitude.

Such a 5 percent GNP loss would be a permanent reduction. That is, at a growth rate of 2 or 3 percent per year, it would be equivalent to something like a two-year setback in the development of GNP after the year 2000. In any year after 2000 the GNP might be 5 percent below what it could have been had the real costs of fuel not doubled. Alternatively stated, from about the year 2000, any given level of GNP would be reached about two years later than that level would have been reached had the average cost of fuel stayed at the level of the late 1970s.

This estimate, based on purely arithmetical operations with a conservative price elasticity, a historically observed GNP elasticity of demand for fuels, and an estimate that the costs of all fuels could on average double again within two decades, is probably pessimistic in purely economic terms. A major uncertainty in the cost of domestic fuel will be the costs we choose to incur to avoid environmental damage, especially the hazards to health in burning increasing quantities of fuel. But the purpose here is only to arrive at a rough estimate of what may be in store, especially if some of the more optimistic estimates of what nature has hidden for us under the earth's surface should be disappointed. What we get is an estimated burden equivalent to a deadweight tax of up to 5 percent on our GNP in perpetuity, or equivalently, a leftward displacement of the GNP growth curve by a couple of years, from and after about the year 2000.

Absolutely, the loss is huge. Five percent of GNP today is about a hundred billion dollars. When GNP has doubled it will be two hundred billion. But in the first few years of the next century it would be two hundred billion dollars subtracted from a GNP of four trillion.

This figure is both immense and modest. If we were calculating the worth of averting a loss of that magnitude, it is an enormous amount of money, equivalent in percent of today's GNP to most of the defense budget or two-thirds of the total outlay for personal health services. But its historical significance can be appreciated by drawing that curve projecting the GNP from now until 2025 on an ordinary printed page; the difference between real GNP with doubled fuel costs and real GNP with today's fuel costs—that is, with the added cost of energy treated as a net subtraction from GNP—is not much more than the thickness of a line drawn with a soft pencil.

This does not belittle the problem. A lot of money is covered by the thickness of the pencil by the time we reach a GNP of four trillion dollars. The problem is a major one. It ranks with, not necessarily above or below, several major foreseeable economic difficulties.

GROWTH, STABILITY,
AND INFLATION

The foregoing assessment considers only the aggregate cost to the nation as a whole of diverting resources into higher-cost energy production or into less energy-intensive consumption. It is an assessment of the total *lost real income*. There are at least two other questions about the way the economy works that have to be examined. One is: What would be the impact of another doubling of fuel prices on overall economic performance? Would the induced conservation of energy use resulting from the doubled price so impair economic performance that some further loss must be accounted for, some multiplier effect on production and employment? Our assumed GNP growth, without fuel price increases, would raise energy use by some 80 percent over the rest of this century. In considering a possible doubling of fuel costs we have allowed for a price response that, over the same period, inhibits some of that increase in the use of energy, so that energy use at the end of the century would be 50 percent rather than 80 percent greater than now (i.e., 15/18 or 5/6 of what it would be at unchanged prices). Should we expect that price-induced reductions of energy use, accumulating over a twenty-year period to as much as 15 or 20 percent, will do disproportionate harm to employment, productivity, and economic growth? That is, in avoiding the full cost increase on the quantities of energy that would have been used in the absence of any price-induced reduction, will consumers and businesses do the economy more harm than if they paid the price increase without economizing in the use of energy?

Another question is: how will the impact of a doubling of fuel costs be distributed through the various sectors of the economy? Will the impact of these costs, or of this lost productivity, be spread over the economy in such a way that costs are shared throughout the population, or instead be concentrated by region, by economic sector, or by income class?

Energy permeates the economy. Pure energy in heat, light, and motor fuels is used by everyone. Some production processes and some consumer goods are much more energy intensive than others, but not many sizeable industries are so concentrated and so energy intensive as to generate isolated serious pockets of depression if the prices of fuels double again. General economic reasoning (as well as some elaborate econometric modelling) finds no reason to believe that in the long run, with steadily rising fuel prices that double in the course of a decade or two, the economy cannot take it in stride. The experience of World War II was that even far more severe short-run dislocations are not a threat to the viability of an economy or to its capacity to remain fully employed.

This conclusion has to be qualified with respect to the balance of payments. During the next decade or two, American oil imports might be anywhere from a low figure of six or eight million barrels per day to a high figure from twelve to fifteen million. Using an intermediate figure of ten million for illustration, and the current price around $14 a barrel, the annual value of oil imports would be $50 billion. If the price of OPEC oil continues to be politically determined, by concerted action among suppliers, there is no assurance that prices will not occasionally change abruptly. The motives could be political or commercial. A sudden increase by, say, 50 percent would have three kinds of effect, and they need to be carefully distinguished.

One effect would be on the users of fuel, the price of whose fuel would increase abruptly depending on how much federal regulation held down the price of domestic fuels. The abruptness and unexpectedness of such a substantial price increase would make accommodation to the higher prices more costly than had they been gradual and anticipated. But the extra costs of adaptation due to the abruptness itself would be short-lived and not cumulatively large compared to the cost increase itself.

The second effect would be an inflationary impulse. A sudden cost increase equivalent to one percent of GNP, together with the associated increases in the prices of domestic fuels, would show up promptly in production costs and in the consumer price index. An economy that faces the chronic danger of general price inflation is vulnerable to any imported commodity whose price can suddenly and disruptively increase to the extent of a whole percentage point in the consumer price index. The impact is greater than the once-for-all change in fuel prices because of the many wage agreements and other cost elements that have a contractual or statutory relation to the price index and induce further price escalation. Offsetting deflationary policies could be triggered that would have a depressing effect on production and employment. How serious those would be depends on how effectively inflation is combatted; but for at least a brief period, there could be a temporary loss of national income beyond the $25 billion annual added cost of foreign oil.

A third effect is the deflationary impact of the immediate shift in the balance of payments. Twenty-five billion dollars of current expenditure would be largely diverted from other consumption into the higher cost of imported fuel, not instantly matched by a corresponding increase in demand for U.S. exports. The fiscal impact is like that of a tax imposed suddenly on fuel without an immediate corresponding increase in government expenditure. Well-designed fiscal programs need have no difficulty in substantially offsetting this deflationary impact, but well-designed fiscal programs of that magnitude are not always readily available and politically acceptable on short notice. (This fiscal impact is separate from the effect on the value of the dollar in world currency markets, which can be additionally mischievous.)

The magnitude of such a deflationary impact, if offsetting fiscal policies were not to become promptly effective, could be on the order of a percentage point in unemployment for a year or more. And again, this is lost earnings additional to the lost real income due to the higher cost of imported fuel.

The experience of 1974 was an extreme example of these effects on price inflation and demand deflation. Except briefly during the period of informal motor fuel rationing, higher fuel prices per se had little or no effect on production and employment. But there was a balance of payments effect, which did aggravate unemployment, and a stimulus to price inflation that severely inhibited the government's fiscal action to offset the impact on employment.

WHO GAINS?
WHO LOSES?

The distributive question—who is hardest hit by increased fuel prices and who least hard, who gains and who loses— has a short-run and a longer-run perspective. The short run relates to the current regulations on petroleum and natural gas, regulations that have kept the prices of petroleum products and gas below the market level. The prices on "old oil"—oil from wells that were fully operating before 1973—have been held to less than one-half the recent OPEC price. Wells brought into production more recently are allowed to sell at a price closer to, but still below, imported oil. Because refineries buy old, new, and imported oil in different proportions, they pay different average prices; under a system known as "entitlements," refineries that obtain a cheaper mix make reimbursement payments, and refineries that use a more expensive mix receive reimbursement payments, with the effect that both pay the same average price. That average price has been about 15 percent below the world oil price. Natural gas shipped across state lines has been regulated at a price that may average about one-half of what it would sell for if gas and petroleum were not regulated.

"Old" oil and gas are a depleting resource. Unless the wells more recently developed, or wells yet to be developed, are legally declared "old" in relation to future increases in world oil prices, the old oil and gas will cease to be a significant part of the total in about six years. Meanwhile, the quantitative effect of price regulation averages about $4 per barrel on some nine million barrels per day of domestic production, or about $35 million per day, $13 or $14 billion per year. For natural gas the figure would be similar. A total in the neighborhood of $25 billion per year is probably the difference between what all consumers, individuals and businesses, pay for their fuel today and what they would pay in the absence of price regulation.

That is a very gross estimate of the "income transfer" from consumers to oil and gas producers, before corporate and personal income taxes and before capture of any of those proceeds in higher wages or absorption in higher production costs. (Some part of the increase in wellhead prices of oil may reduce the refinery profits.) Some fraction of the dividends and capital gains arising from price deregulation would accrue to pension funds, insurance companies, and the like. As a very crude estimate we take $15 to 20 billion per year as the likely current net redistributive effect of recourse to free market prices; this figure diminishes each year as "old wells" are exhausted.

This $15 to 20 billion per year in net redistributive shift from the rest of us to the owners of petroleum and natural gas resources is primarily what the present policy debate is about. There are two sides to the debate. Should *consumers pay* prices for oil and gas that reflect their market values and the estimated costs of replacing gas and oil with future production? And, if so, should *producers receive*, on oil and gas from wells, many of which were brought into production at much lower market prices, the full proceeds that would accrue from consumers' paying market prices?

The central current issue in the policy debate is whether consumers should pay up to $20 billion a year more for gas and oil and, if so, how much, if any, of the proceeds should go to producers of oil and gas, and what should be done with the difference. The issue is divisive, and there are regional and other differences in impact to heighten controversy. The amount is as big as most controversial economic figures that arise in a single year; it is equivalent to major tax reform or aid to the cities.

It is not an amount that staggers an economy, reverses historical trends, or changes the quality of life or the character of society. The impact on the poor is somewhat, but only somewhat, more than in proportion to their share of income; the effect on consumers would be about like a transient 2 percent sales tax in its magnitude and incidence.

In the longer run, the issue of what to do about the pricing of "old oil" and "old gas" will persist only if regulation is indefinitely continued. Price regulation is a *distributive* issue for the simple reason that it cannot keep down the costs of fuel, it only determines who pays them.

FACING UP TO THE TRUE COSTS

Our estimate was that the costs of fuels are likely to increase over the next twenty to twenty-five years and that the increases, though substantial, are unlikely to exceed the equivalent of 5 percent of the GNP in added cost or lost income. That estimate we said was probably pessimistic *in purely economic terms*. In political terms it could prove optimistic. The estimate relates to the costs that may be unavoidable, the costs that are determined by technology, geology, demography and economics. Decently managed, the energy component of our economy need not be expected to interfere seriously with employment and continual economic growth and it need not entail costs of a magnitude to deserve much attention from economic historians in the future. But that estimate did not include an allowance for mismanagement.

The danger is that we shall attempt to insulate ourselves from the rising costs of energy, deceiving ourselves that because we do not pay the costs directly they do not have to be paid.

Energy policy itself can aggravate the problem by dealing superficially with its manifestation, by attempting to hold down prices while genuine costs are rising. If the true costs are not faced we shall waste our energy resources in consumption, deny ourselves the enlarged resources that would be available at higher prices, and delay the technological changes that higher costs would encourage.

3

THE CRITICAL ROLE
OF OIL IMPORTS

Oil imports, it was noted above, are the critical connection between energy in this country and in the rest of the world. They are the main and virtually the only channel through which world energy supply and demand currently impinge on domestic U.S. energy. They are the main way, although not the only way, U.S. energy supply and demand impinge on world energy. They are the principal energy connection between the United States and a multitude of energy-related strategic and foreign-policy issues.

U.S. imports of oil therefore deserve particular attention in the design of energy policy. The President's National Energy Plan of 1977 made the level of imports a central target. They should be a central concern. But the fact of their importance does not determine what our policy should be. The issues are as complex as they are important.

In the first place, the many international issues that are wrapped up in the world energy problem—NATO and Japanese security, east-west relations, the danger of war in the Middle East, development of the Third World, cooperation on nuclear-materials security—will not be managed mainly by oil imports. In coping with those issues it will be alliance policy, trade policy, nonproliferation policy, arms-sales policy, Soviet-American relations, and a multitude of other foreign-policy dimensions that will principally determine success or failure.

Even relations with Iran and Saudi Arabia depend on more than oil policy.

The world's fuel problems, not to mention the surrounding political and strategic problems, cannot be decisively moderated by U.S. policy on imports. The world's fuel problem is large, permanent, politically and technologically complex, and mostly beyond American control. Turning down the faucet on U.S. imports will not take care of it.

Nevertheless, a difference of five or six million barrels of oil per day, between nine or ten million and fifteen million barrels in the 1980s, would be a large difference. At the current price of oil it would mean a difference in our international payments of $80 billion instead of $50 billion per year. If the five or six million barrels of additional imports induced a difference in the price of oil by, say, $3 per barrel, it would mean $100 billion instead of $50 billion in our annual payments. Such a price difference would also add a like amount to the costs of other countries' oil imports. Policy on imports is therefore a big part of any effective approach to world energy.

A policy that significantly moderates our need for oil is a signal to other countries that we can keep our balance of payments under control. It is a signal that we will moderate not only our demand on world supply but our upward pressure on oil prices, helping other countries with their payments' problems as well. Furthermore, an energy program determined to keep oil imports within reasonable limits may provide some leverage on the oil-import policies of other countries. The level of other countries' imports will be determined, in part at least, by the prices we pay for ours. Using the bargaining power that would accrue to us from our willingness and ability to hold imports within agreed limits will add a crucial and difficult dimension to our oil policy.

OTHER CONNECTIONS

Before pursuing the implications of the central role of oil imports, it will be useful to avoid over-simplification by noting other energy connections.

A major one is international trade in *nuclear fuels, reactors, technology, and waste products.* Nuclear power will become a major world energy source during the next two or three decades, and its foreign-policy and national-security significance is beyond exaggeration. As CED's policy statement *Nuclear Energy and National Security*[1] made clear, U.S. domestic nuclear-power decisions will not have a decisive impact on the programs of foreign countries. U.S. concerns relate mainly to the proliferation of weapons material and technology; and the U.S. means of influencing those developments are not tightly connected to the way nuclear electricity is developed or regulated in the United States. Nevertheless this is an important connection between U.S. and world energy. An assured adequate supply of low-enriched uranium reactor fuel can be far more important in helping to meet non-proliferation objectives than its export value would indicate.

Nuclear power is, of course, an area in which formal international cooperation plays an important role. The United States is taking a leading role in the examination by some 40 governments of various ways to assure security of fuel supply and reactor waste disposal for the countries that now have or presently will have nuclear electric power programs. Assuring an adequate supply of reactor-fuel for other countries, and an adequate U.S. capacity to provide low-enriched uranium fuel are simultaneously substantial contributions to world energy and potentially major contributions to security against the proliferation of weapons-grade nuclear materials. They are also modest contributions to our energy balance of payments.

A second connection is *research and development.* The United States shares technical leadership in energy research and development with a number of countries and is by no means a unique source of the world's future technologies. But many of the technologies under development in this country could be important to energy supply or conservation in other countries. There is scope for making U.S. research and development more

---

[1] *Nuclear Energy and National Security*, a Statement on National Policy by the Research and Policy Committee of the Committee for Economic Development, New York, New York, September 1976.

responsive to the needs, resources, and opportunities of other parts of the world, especially the developing countries.

Coal (except metallurgical coal) is not presently a major U.S. export. If the price of oil should resume a steep uptrend, U.S. coal transport to Japan or to Western Europe might become commercially attractive. Export of a hundred million tons or more per year, at a value of five to ten billion dollars, is not out of the question. But unprocessed coal is used in electric power production, and nuclear power is likely to appear a cheaper and environmentally cleaner substitute, even a more secure supply, in the larger countries that might consider importing coal. Converted to liquid, coal will not be competitive on a large scale until, at the earliest, two decades from now; coal thus used would help to reduce oil imports. Coal production will be environmentally limited, and a large exportable surplus may depend on more rapid expansion of mining and transport than can occur. Nevertheless, in the longer run of two or three decades, U.S. coal exports could become significant.

There is one potentially significant *environmental* area in which U.S. and worldwide uses of energy join to produce a common problem. It is the accumulation of carbon dioxide in the atmosphere, a problem only beginning to be systematically explored. It is potentially awesome in the effects on climate that are considered possible, but uncertain, poorly understood, and probably some decades away from requiring action.

Finally there is the entire systemic relation between U.S. economic stability and growth on the one hand, and the economic health of the rest of the world on the other. The volume of U.S. imports and exports, the volume and direction of capital flows, the institutional rules governing trade relations, and a number of areas of international cooperation and foreign aid, are economically and politically affected by the health of the U.S. economy—our growth rate, our inflation, our foreign-currency rates of exchange, our investment markets, and our ability to avoid protectionist depressed areas. There is a linkage between our energy policies and the health and growth of the U.S. economy, a linkage that will be the more noticeable the less well we manage our energy problems. There is linkage between the U.S. economy and the economies of the rest of the world. And there is linkage between those economies and the "energy problems" embedded in them. But the specific character of any U.S. domestic energy event, in its ultimate impact on energy events in the rest of the world, will be hidden in the complex transmission of overall economic performance, except as it relates directly to the volume of oil imports.

Oil supplies from OPEC countries may be partly dependent on the willingness of those countries to invest in the United States. That in turn will be affected by their perception of the health and growth of the U.S. economy and the stability of our financial institutions. If mismanagement of our energy problems appears to jeopardize the U.S. economy or the U.S. balance of payments, the availability of world oil could be diminished.

THE WORLD OIL MARKET AND THE WORLD OIL PRICE

The President's National Energy Plan included a prospective reduction of oil imports by 1985 to barely half of what they might have been by then in the absence of import-reducing policies. Accomplishing that would require large reductions in energy use, substitution of other fuels, or increases in domestic oil supply. Three issues are involved in devising a policy for oil imports. One is the *quantities*, and whether the quantities should be flexible targets or firm programs. A second is the *program techniques* by which imports would be made to conform to some targets; these could range all the way from fees or direct controls on imports themselves to subsidies for home insulation. A third is the *diplomacy*—the negotiations, commitments, and the cooperative arrangements—with which our goals and programs are worked out or discussed with the producing countries and the other consuming countries.

In addressing these questions it is worthwhile to review some of the roles played by oil imports and the nature of the world oil market.

The market for oil is substantially a "world market." Shipping and marketing of crude oil and refinery products are highly decentralized. End-use controls on refinery products are nearly impossible. Oil exports can be shifted in destination; transshipment of oil and refinery products is easy. The oil producing countries lack any policing mechanism to preserve discipline on selective import controls. Denying a large fraction of world oil to particular countries may not make much difference if the remaining fraction is adequate to their needs and is available at no great change in price.

The implication of this is not only that an embargo of oil is a diffuse and inaccurate weapon, but that consuming countries share a common problem whenever oil exports are interfered with. The severity and timing of the problem would differ among countries in the event of embargo or obstruction. But what one country can do with strategic stockpiles or with emergency conservation will be of interest to the other countries. And what a large consumer like the United States does in an emergency will be perceived either as a major contribution to international cooperation or as a major subtraction from it.

As long as we take seriously that our European allies, Japan, and other countries are part of our mutual security system; as long as we care about French or Japanese policy toward the Middle East; as long as we care about successful development in the poorer countries of the world, the most serious "vulnerability" of the United States to a contrived energy emergency is likely to be the effect it will have on other countries that matter to us, even more than its effect on us.

Among the important characteristics of imported oil in the U.S. economy, the first to note is that it is a major source of energy. It is one-fifth of all the fuel we consume and currently it is not getting smaller. Furthermore it is a flexible source, increasing or decreasing easily on short notice, cushioning the excesses and shortfalls of different supplies and different demands—a cold winter, a coal strike, delays in nuclear power plant construction, or shortfalls in conservation policy. In the absence of disrupting influences, the world supply is a great reservoir for cushioning the vagaries of supply and demand and government policy.

Consuming imports saves domestic reserves. At the same

time, in the absence of measures to develop unused capacity, consuming foreign oil instead of producing domestic oil diminishes short-run domestic productive capacity because of the time lag in bringing in new wells or enhancing output from old wells and enlarging distribution facilities. There is some trade-off between domestic reserves and domestic productive capacity: the more we import the more we save; the more we import the less we currently produce; the less we currently produce the more vulnerable we may be in the short run and the less vulnerable in the longer run.

Oil can be stockpiled as a strategic reserve against interruption in supply. Member countries of the International Energy Agency [IEA] have agreed to establish, by 1980, an emergency reserve equal to ninety days of the previous year's imports. The current U.S. program is to reach a government reserve of at least a billion barrels by 1985, equivalent to at least four months' imports at the 1978 level. How long such a reserve would last would depend on the severity of reduction in availability and the effectiveness of measures to conserve oil. Even if import supplies were reduced by 25 percent, a 10 percent reduction in import consumption (corresponding to reduced consumption of oil by no more than 5 percent, less if production at home can expand promptly) would draw down reserves by only 15 percent of the normal rate of imports, and a reserve equivalent to four months of normal imports would last more than twenty-four months. Of course, stockpiling does increase the current demand for oil in the short run; whether the supply situation will ease enough to permit some accumulation of reserves during the next few years without undue upward pressure on prices will depend crucially on events in Iran.

We are ourselves vulnerable to the needs of other consuming countries in that our foreign policy objectives can oblige us to restrict oil imports somewhat painfully to help make oil available to other countries whose claims to a reasonable share we must recognize. It is important not only that we help ease the oil import and price problems of other countries, but that we appear cooperative and exercise leadership in the interest of alliance relations, nuclear non-proliferation, and the other key objectives that are intertwined with energy. The United States has agreed to share available oil resources with the other IEA governments and to draw down emergency stocks in the event another OPEC embargo causes shortfalls that exceed specified percentages of normal supply. Performance in such an emergency is not guaranteed, but the leadership and leverage that the United States could exercise would depend not only on the willingness but on the ability of the United States to limit imports in accordance with the sharing formula in such an emergency.

Finally, the relation of oil prices to the quantities we import, or to the quantities that other countries import, can be crucial in determining the "cost" of additional imports or the "savings" due to reduced imports. There is no reliable way to calculate the effect of changes in world import demand on the prices that will be set or obtained by OPEC countries. It is almost certain, however, that reduced demand for imports will soften prices or slow their climb, while increased demand would hold prices up or accelerate their climb, especially when, as it is expected, total demand for OPEC oil approaches the projected limits on OPEC supply capabilities in another decade. A consequence of this price-quantity relation is that the true economic cost of importing more (or the costs saved by importing fewer barrels of oil) is not accurately measured by the price per barrel at which the oil is imported.

Purely as illustration, consider the difference between the estimate in the National Energy Plan that its proposed program could result in oil imports in 1985 of about seven million barrels per day compared with twelve to sixteen million in the absence of the program. Scale that import difference down to a more modest four million barrels per day, eight million vs. twelve million barrels. Assume the price corresponding to eight million barrels would be $15 per barrel, and that with the higher level of imports (perhaps with some imitation by other IEA countries that observe the U.S. not greatly conserving imports) prices in the late 1980s would be higher by 20 percent, or $18 per barrel. The "cost" to American consumers of the additional four million barrels is not just the higher price of $18. Eight million barrels at $15 would cost $120 million; twelve million at $18 would cost $216 million. The difference is $96 million, which comes to $24 per barrel for the extra four million. (The four million additional barrels cost $18 apiece and add $3 per barrel to the eight million barrels already being imported).

(Moreover, other countries also are paying an increase of $3 per barrel as a cost of *our* extra four million barrels.)

In the same way, if we hold imports to the lower level but other consuming countries let their imports increase so much that the price is $3 higher after 1985, we pay the extra $3 on our eight million barrels, or $24 million per day as *our* "cost" of *their* additional imports.

A final point needs to be made about the market for oil. We noted above that it is indeed a "world market" because of the large quantities that can shift from particular buyers to particular sellers, the ease with which oil can be redirected or resold, and the impossibility of identifying individual refineries' products and policing any program of selective denial. That is quite separate from whether competition determines the market price or prices are set arbitrarily by a cartel, or at least substantially influenced by a single large supplier or a few suppliers that together can manipulate the price by the quantities they are willing to sell. It has been argued that OPEC is by itself responsible for the energy crisis and that a primary policy objective should be to find a way to dissolve OPEC or to coerce it, through economic measures or otherwise, to lower its price. The price of petroleum advanced so decisively when OPEC acquired political cohesion in the aftermath of the October War of 1973, leading to the popular discovery of the "energy crisis," that there is a temptation to believe that if their political cohesion—initially an Arab-nation phenomenon arising in war —could be dissolved or moderated, petroleum prices would recede and the energy crisis would return to its benign nonexistence of a decade ago.

Looking backwards five years, one may find that interpretation of the "crisis" plausible. Looking forward any distance, the coming increase in the cost of fuel appears to be much less dependent on OPEC behavior. Although the oil-production policy of Saudi Arabia could indeed make a difference of several million barrels per day in the late 1980s and the 1990s, it would only moderate the upward pressure on oil prices by depleting reserves more rapidly.

The diagnosis that the energy crisis is purely a cartel phenomenon is contradicted by two considerations. First, even with substantially expanding OPEC production over the next two decades, it is unlikely that petroleum prices can decline except occasionally for brief periods, and likely that the net change will be substantially upward. Second, oil in the ground at today's prices is not an unreasonable investment, especially for Saudi Arabia that has large liquid financial reserves earning a nominal rate of return not much better than the rate of inflation, considering the rate at which petroleum prices may rise during the next decade or two. As long as oil revenues to Persian Gulf nations exceed their current import requirements, expanded production converts oil in the ground into alternative financial or physical assets. The oil-producing states used to be desperate for foreign exchange and liquid assets, but now they can afford to hold oil rather than pump larger quantities at lower prices. It is likely, therefore, that the discipline of an OPEC cartel has been superseded by national self-interest as a motive for not maximizing production in the short run, at least for Saudi Arabia and some other Persian Gulf countries.

Clearly, the world energy problem described in Chapter 1 is not merely the artificial construct of a cartel. Nor is it certain that the situation over the long run would be greatly improved if OPEC countries dumped substantially larger quantities of oil onto the market, greatly depressing current prices while depleting their reserves more rapidly.

In sum, espousing a policy of "breaking the OPEC cartel" could be self-defeating on three grounds. First, if it succeeded it would not make the energy problem go away. Second, it would be economically mischievous if it encouraged the fantasy that the energy problem is going to disappear and prices are going down. Finally, its diplomatic effect would give credence to the most immoderate participants in OPEC.

REDUCING DEPENDENCE ON IMPORTS: THE POLICY CHOICES

It is characteristic of the list of domestic measures that appears on page 9, and of those proposed in the 1977 National Energy Plan, that none would have a *prompt* and *substantial* effect of *predictable* magnitude on the level of imports. While we can usually be sure of the direction of impact, nobody can estimate the speed and magnitude of the effect.

Studies by the Congressional Budget Office and by the General Accounting Office disagreed with the estimates in the National Energy Plan, and the methodologies of those two critiques show how easy it is to be off by a couple of million barrels a day in estimating petroleum imports, even in comparatively straightforward proposals, to say nothing of the exact size and timing of effects that changes in strip-mine legislation or offshore leasing policies would have.

A major reason for this imprecision is that, under present policies, imports are a cushion. If the programs adopted fail to reduce the excess of U.S. demand over U.S. supply on schedule, imports fill the gap. They moderate the domestic shortages or price increases that would result if imports were unobtainable.

Therefore, if a multitude of measures proves inadequate to the reduction of imports in conformity with some program goal, new methods have to be instituted and we have to wait and see how they work in reducing the demand for imports. The only measure in the National Energy Plan that was directly geared to the amount of petroleum consumption, and hence related directly to the volume of oil imports, was the threatened gasoline tax that would have been introduced at the rate of 5¢ per year starting in 1980 if the gasoline-consumption goals were not being met. Even that tax—which Congress did not seriously consider—would have had an impact delayed in years, not months, and an impact of unpredictable size.

If a reduction in our *dependence* on oil imports—whether an actual reduction in imports, a levelling off, a slower rate of growth, or response to emergency—is considered important for foreign policy, balance of payments, vulnerability to disruption, or orderly progress toward lower oil imports in the future, there will have to be policies that work more effectively on imports than those in place today. "More effectively" can mean more *sizeably*, more *predictably*, more *promptly*, more *sustainably* (i.e., with less troublesome side effects), or any combination of these.

If our main purpose is reducing vulnerability to disrupted supply, the emphasis should be on promptness and predictability of the measures put in place. If the main purpose is moderating world prices in the 1980s, size and predictability, or a combination of prompt effect and flexibility of administration (to permit successful trial and adjustment) will be important. If the primary goals of import moderation are longer term, then size and sustainability of policy effects need to be emphasized along with the incentives that any such measures provide for bringing in new supplies and new technology.

This choice of the focus of emphasis should, in turn, condition the choice of measures used to reduce import dependence. The relative importance attached to the different objectives should broadly determine the choice between those measures that reduce import demand *indirectly*—inhibiting demand, encouraging conservation, stimulating supply, and inducing technological change—and more *direct* actions to inhibit or restrict imports. Among the former would be imposing taxes or offering subsidies or tax relief, or regulations, that discourage the consumption of fuels, especially liquid fuels. It would include efforts to reduce environmental obstacles and delays to enlarged production. It would particularly apply to policies that would allow domestic energy prices to rise fairly promptly to world levels. The more direct actions would be fees or import charges, licensing of imports, auctioning import quotas, or cooperative arrangements among oil importers.

In general, direct means of intervention can be more prompt and predictable in their effects, if there is authorization for them. Obtaining new authorization can be a source of delay (as the congressional response to the original National Energy Plan dramatically illustrates). The former, indirect means, can eventually have sizeable if slower effects and fewer undesired side effects.

The "side effects" are no mere academic detail. There is no way that direct action against oil imports can be separated or insulated from domestic energy prices. Any program that directly inhibits or regulates imports will pose hard choices in domestic fuel price and regulatory policy. This can most readily be appreciated by considering what ought to be the first and most obvious action to reduce imports: to let the price paid by consumers of fuel rise to equal the amount actually paid by consumers collectively for the oil that is imported. Consumers presently pay about $2 per barrel less for imported oil (and for all oil) than the import price. The importing refinery or distributor pays about $2 less per barrel, subsidized by the averaging process that withholds from domestic producers part of the price consumers pay and uses that difference to hold down the price of imports. Every barrel of imported oil costs consumers collectively the full OPEC price, but the consumer shifts $2 of that price to other consumers by raising the average price.

Making consumers pay the full price of imported oil is an obvious first step in rationalizing our fuel policy. The 1977 National Energy Plan provided this, in a succession of annual steps. The effect of an increase by 15 percent or so in the price of imported oil might not be substantial, but it would be at least in the right direction and in the longer run would induce better decisions on supply, conservation, and technological change.

More controversial is the unavoidable domestic counterpart to that decision. Imported oil looks like, burns like, and is like, domestic oil. There is no sensible way to charge people more for gasoline refined from Arabian or Venezuelan oil than for gasoline from Alaskan or Gulf States oil. Letting the consumer pay $2 more per barrel on eight or nine million barrels of imported oil entails letting the consumer pay $2 more per barrel for domestic oil as well. This is where the choice gets hard and the controversy bitter.

We are back at the distributive issue discussed earlier. The extra $2 paid by the consumer for imported oil does not go to the foreign supplier; the foreign supplier has always been receiving the world price. The issue is whether domestic producers should receive the world price for their oil, or should some kind of wellhead or other tax be imposed to keep the full price increment from being a "windfall" for domestic producers. This is the issue that was discussed at the close of the preceding chapter.

Now consider a proposal to go even further, reducing oil imports below the level that would correspond to the equalization of domestic consumer prices with world prices. Broadly speaking, there are two ways to do this: with price regulation, and without it. If prices are regulated while imports are reduced, domestic shortages will appear that require some kind of allocation or rationing, or direct measures that reduce demand (equivalent to rationing) that work outside the price system. Permanent rationing and price control are "side effects" of sufficient potential cost to weigh heavily against the benefits from reduced imports.

If price controls are eschewed, the price of oil will have to rise sufficiently to bring about the programmed reduction in demand for imports. The domestic price must rise above the world price of oil, and a duty or import fee of some kind must be added on top of the world price of oil. Domestic prices will rise not merely to the world price but to the world price plus that duty.

There are two sizeable consequences of this approach. First, the "distributive problem" emerges again. An import duty of $2 or $3 per barrel transfers income from oil consumers to the federal treasury where, however bitter the partisan dispute about the distribution of those proceeds, it is available for some kind of redistribution. Domestic petroleum, selling at the same price and not taxed, yields proceeds not for the federal treasury but to the producers of oil. That aggravates the distributive issue. If the domestic output is taxed like imports, the desired effect of the price increase on domestic supply will be depressed.

Furthermore, the effect on the consumer price index, and on all the wage and price decisions and agreements that are institutionally related to that index, are likely to be the same as if the price of oil increased on its own, and not as a result of incentive taxes on oil. An extra $2 per barrel on imported and domestic oil—a fairly moderate inhibitory tax incentive from the point of view of energy policy—would add a percentage point or two to the consumer price index; and while its impact need not be altogether intimidating, it is far from negligible.

That the choices involving direct action on oil imports are hard ones does not mean that the more indirect measures, those that might act more slowly over time, are easy ones. Whether school buildings or homes or government offices are insulated because of the high price of heating oil, or because of regulations that make it mandatory, the insulation has to be paid for. The point is not that direct action on imports will have side effects and the indirect actions will not, but that the differences among side effects of different kinds of import-conserving policies are so important that they cannot be disregarded in the evaluation of costs and benefits.

4

THE PRICE SYSTEM AND ITS LIMITS

This section recapitulates the themes and observations of the preceding sections and formulates principles that should govern policy. It will not propose specific policies. Many alternative combinations of programs could be responsive to the same needs and principles. They would differ in detail—in the way that benefits and burdens are distributed by region, by industry, or by income level; in their reliance on prices and other economic incentives; in the level of government or the agency of government that formulates or implements the programs; in the year-to-year or region-to-region flexibility with which they may be administered; in the visibility of their application and their effects, and in the awareness of consumers, workers, or property owners of their incidence.

Furthermore, policies should be consistent; the whole program should have a certain balance. Principles can conflict, and the conflict among them should not be disguised. It is in the formulation of specific policies that the compromising, the balancing, and the marginal adjusting should be done to minimize the conflict in the interest of fairness and acceptability.

The preceding chapter emphasized the uniqueness of the connection between the domestic energy economy and the world energy economy, the oil-import connection. A contrasting observation is fundamental to domestic policy. It is that the interconnections within the United States among the different fuels, the different uses of fuels, and the different users of fuels, are multifarious.

While it is not true that coal can substitute for gas in all its uses or that refineries can shift their output mix to 100 percent gasoline or home heating oil, or that we can bake waffles on coal stoves or light our dining rooms with natural gas again, it is true that there are wide margins for making inter-fuel substitutions and for substituting materials, activities, and consumer goods that differ in their demands for energy. Natural gas is simultaneously used for industrial heat, electric power and home heating, and it is easily transported hundreds of miles. Coal, oil and gas can all be used for electric power and are all used for industrial heat. Reduced gasoline consumption allows more home heating oil to be produced from the crude, alleviating the pressure on natural gas or nuclear electric power.

That is why there are so many different ways to devise an energy-policy package. It is also why energy policy is bound to be controversial. If there were but one thing to do, as there occasionally is when an electric grid is over-burdened on a summer afternoon and power cutbacks are unavoidable, there would be little to argue about except the level of risk we were willing to incur, or the investment we would make today to ease the problems tomorrow. It is because there are so many ways that energy can be saved, rerouted and allocated, its production subsidized or its use penalized; so many ways of financing the subsidies or utilizing the tax proceeds; so many ways of discriminating between homeowners and apartment dwellers, the elderly who need warm houses and the motorists who need gasoline, industries in New England or industries in the North Central states, producers of oil from old wells and explorers for new wells—that any major policies affecting this hundred billion dollar component of the GNP generate conflicts.

PERSPECTIVES IN TIME

As we scan the coming century we see a succession of loosely defined and overlapping periods that correspond to changing energy objectives.

There is an *immediate period*, less than a decade, in which limits on domestic fuel production have to be taken almost as fixed. Fuel consumption has only begun to respond to the price increases of the 1970s; and the dominant policy issue is whether consumers should continue to buy petroleum products and natural gas at prices below the world price for oil and the equivalent market value of gas. As "old wells" are depleted, the dollar value of this issue will decrease but not disappear.

Isolated from the future and from the rest of the world, this issue appears to be purely distributive. But the future is not isolated. Future production depends on decisions made in this immediate period on the basis of prices *expected* during the second decade from now. Similarly during this period we are not insulated from the rest of the world. The demand for imported oil reflects the below-market price that refiners and consumers pay. Our balance of payments reflects this subsidy. Thus what appears to be a purely distributive issue in the short run is a long-term supply issue, a long-term conservation issue, and an immediate as well as long-term balance of payments issue, because it cannot be detached from the longer future and the wider world.

A second time perspective is *from now to the end of the century*. During that time most of our energy will continue to come from oil, gas and coal. Increasingly nuclear power, now not quite 10 percent of our electricity, will replace the fossil fuels. But by the year 2000, nuclear power actually on line will be confined to the plants initiated during the coming dozen years, and even if all electricity growth were in nuclear plants from now on, the fraction so fueled by 2000 will not exceed one-quarter. Solar heating, especially space and home water heating, can be increasingly installed and even some start on solar-powered electricity may appear.

Liquid fuel from shale and other unconventional sources will entail not only further development but eventually large investments in new kinds of extracting and refining equipment, and environmental and land use problems that have hardly been investigated, much less resolved. The expected availability and prices of such fuels will have an impact on fossil fuel development in the preceding decade, but actual consumption of such fuels during the remainder of this century will be small. Coal-based gas or liquid fuel could become commercially available before the end of the century, but actual growth of a coal-based synthetic fuels' industry will begin only in the 1990s. The quantities, though noticeable, would not be a significant percentage during any appreciable part of this century. Since nuclear power produces only electricity, the demand for liquid fuels will still have to be met from petroleum.

A third period, less confidently foreseeable, might be from about the *beginning of the next century* and lasting some decades. That will be the period when world production of gas and petroleum may be absolutely declining, despite continually rising demand for energy, and costs are primarily determined by how rapidly and how economically alternative sources of energy, not in use today, can be relied on. The use of sunlight directly for electricity; liquid fuels from coal, shale, tar sands; and perhaps nuclear reactors not dependent on large quantities of uranium, will be competing. Production of liquid hydrogen and even of alcohol could be part of the mix.

Still a fourth perspective, overlapping the third, may have to be located some time in *the next century*. This would be a period in which energy decisions might have to be substantially dominated not by the costs of different kinds of fuel but by the consequences of burning them. If it turns out that the increasing use of fossil fuel jeopardizes climate and productivity, by what the carbon dioxide and other combustion products may do to the thermal properties of the earth's atmosphere or as concentrations of certain elements in air or soil are determined to be too dangerous to health, it could become globally obligatory to reduce our collective burning of fuels around the world. If that time should arrive, or should be foreseen, there will be the economic and technological problem of drastically limiting the worldwide use of fossil fuels and moving toward reliance on solar and nuclear energy. More than that, there will be a planetary political problem in getting acceptance of some scheme of self-imposed global (but not necessarily universal) rationing. That problem could make some of today's issues, like gasoline taxes or deregulation of natural gas, look easy.

THE ROLE OF THE PRICE SYSTEM

In devising energy policies what is by far the single most important principle is also the most controversial and the most misunderstood. It is that producers of fuels and consumers of fuels are guided by prices, current and prospective. If the prices consumers pay reflect the genuine economic costs of the fuels, the fuels will be used only up to the point where their costs are matched by their value to consumers. If the prices anticipated by producers reflect the value to consumers of additional energy supplies, producers can afford to expand production as long as the fuels they produce are worth more than the resources that go into their production.

For all its imperfections the market—when it is allowed to work—is the only comprehensive source of reliable signals to users, savers, and producers, of the value of the energy that, directly or indirectly, they are producing, consuming, or conserving. Market prices provide the *information* by which people can economically adjust their behavior and the *incentive* to do so. If the costs are paid by those who consume the energy, and they know it; if the savings accrue to those who respond to costs in conserving it; if earnings accrue to those who can shift energy to the consumers who will pay more because it is worth more to themselves or to their customers; and if investment in new production or new technology will be profitable when, and only when, the new oil or gas or coal or nuclear or solar energy is worth enough to somebody to cover the full cost of production; business firms, consumers, and government agencies will have their energy activities coordinated by market prices in what is undoubtedly, though far from perfect, the most cost-effective way that can be devised.

The danger, of course, is that we will attempt to insulate ourselves from the rising costs of energy by holding prices down. In that way, we will be deceiving ourselves into believing that the costs do not have to be paid, because we do not pay them directly and openly. But if they are not paid, the energy will not be there when we need it. If the costs do not have to be paid by users, consumers need not care and cannot know what it is worth to save energy.

If an attempt is made to hold prices down while genuine costs are rising, there is the danger that "energy policy" will aggravate the problem it attempts to solve by dealing superficially with its manifestation. If the true costs are not faced we shall simply waste energy resources in consumption, deny ourselves the enlarged supplies that could be available at higher prices, and delay the technological development needed to cope with rising costs.

Opinion polls and congressional behavior indicate that nobody likes prices to go up, certainly not by means of a deliberate policy. Those who believe that there is indeed an energy problem often appear to believe that the problem is to be solved by forcibly bringing demand and supply into alignment, not by letting prices bring them into alignment. Price increases then look like the problem itself, rather than as reflections of the problem and part of the mechanism of solution.

Furthermore there is the dilemma, mentioned earlier, of the desire to redistribute the burden of increasing costs in the short run *and* the need to let expected future prices reflect estimated future costs. Not only do the public and Congress dislike price increases, but many of them can do the arithmetic and understand that a few tens of billions of dollars per year, for at least a few years, would be a net transfer to those who own or have contracted for "old oil" and "old gas". Amounts of money of that magnitude, so easily identified by who gets it, are often considered fair targets for redistribution. That was the basis for the proposed wellhead tax on old oil and the origin of the dispute about what to do with the proceeds of such a tax.

It is easy to make the case that perhaps $15 billion per

year of crude oil revenues are available for the taking by federal tax, with little deleterious effect on the immediate supply of oil (and little net loss to oil producers if the alternative is to retain the current regulated prices). Many congressmen and the administration apparently feel that oil profits have been rendered adequate if not excessive by the political events in the Middle East in 1973, so that in fairness, the higher prices designed for their effect on demand need not benefit the owners of those operating wells. Discriminating in favor of "new oil" and against "old oil" appears to take care of the supply incentive.

But old oil was once new oil. Today's new oil may be declared old tomorrow, and tomorrow's new oil declared old the day after. The same logic by which this year's "windfall gains" can be taxed away while letting consumer prices go to a market level may be just as appealing when oil and gas prices have increased another 20 percent or 50 percent.

The market cannot be divided convincingly between "present" and "future". Its time dimension is continuous. The prices to which today's behavior is a response—the prices that provide the incentives for current decisions on future supply and new investment—are the expected prices for five, ten, or even twenty years from now. Even consumer decisions on heating systems, insulation, or gasoline mileage of the automobiles they buy, depend on the prices anticipated for five years from now. Development now of new fuel sources that may begin to come on the market ten years from now will be a response to the prices expected in the second decade hence. It is *predicted* prices, even more than current prices, that determine investment decisions.

In the best of circumstances there is uncertainty, and no guarantee, that market predictions will be close to the mark. But political predictions by producers and investors are almost certain to depress anticipated prices below what a market analysis would indicate. There is now an impressive record of regulating energy prices to keep them from rising. The National Energy Plan of 1977 proposed a wellhead tax on oil that would allow consumer prices to be based on "the real value of oil" (identified as the "world price of oil"). It proposed a permanent wellhead tax on domestic oil that would keep the net price to producers equal to the 1977 world price, adjusted upward only to keep pace with inflation. That is, in that plan, prospective future oil supplies appeared destined to be worth—to those who found the supplies, developed them, and brought them into production—only what oil was worth on the world market in 1977.

The odds are therefore biased against those who would invest in new sources of energy. If market prices turn out to be unexpectedly low, the investment will be unrewarded; if prices turn out to be unexpectedly high, there is danger of price regulation or taxation. Like the income-tax treatment of gambling, you keep your losses but share your winnings with the Internal Revenue Service. If one accepts the energy projections contained in the National Energy Plan, it is not only in the event of "unexpectedly" high prices that a wellhead or other tax would drive a wedge between the "real value of oil" and the "real worth to the producer" of bringing in new supplies; it is even "expectedly" high prices whose incentive on supply is dampened by the promise of permanent price controls on oil or, equivalently, permanent taxes on it.

The same principle applies to natural gas. Even investments in coal and other energy sources will be influenced by an apparent philosophy of permanent energy price regulation that, originally applied to oil and gas, might be widened in its application.

There is no constitutional way that the government can commit itself to a hands-off policy, or even an evenhanded policy, ten or fifteen years from now. But a rapid and unconditional phasing out of price regulation could make a convincing demonstration that at least one administration and one Congress could agree on a more nearly free-market strategy. Phasing out the regulation slowly, "painlessly," and with reservations, can be unconvincing.

There is no effective way to keep today's policy on the treatment of old oil and old gas from casting a shadow on the future. The immediate value of holding down oil and gas prices for the consumer, or of letting them go up but recapturing the difference in a tax that can be used to provide relief in other forms, is substantial; but it is at the cost of permanent aggravation of an energy problem that will be serious enough even with price policies that stimulate appropriate responses.

If indeed, as the President said in April 1977, the energy crisis is a great challenge, facing that challenge will mean resisting the temptation to evade the very price responses on which our future supplies will depend.

THE MISLEADING IMAGE OF THE "GAP"

This view of the future contrasts with most official analyses. We do not focus on an event expected some time in the late 1980s or early 1990s—the overtaking of world productive capacity by world demand and the emergence of a gap or shortfall.

A real gap, an observable one that actually occurs, would have to result from price control. A gap can occur if prices are controlled below market levels; but unless it is associated with an effective rationing system, there will usually be some kind of cost—time spent waiting, risks of non-delivery, barter arrangements, even bribes—that supplements the price as a moderator of demand.

Usually the "gap" is an analytical construct. It is an estimate of what production would be, and what demand would be, at some specified world price, with either prices unchanged from current levels or prices higher by some hypothetical increment. This construct is intended to illustrate the supply-demand relation of the future on the simplified assumption of unchanging prices or of prices moving in some specified and easily comprehensible path. It simplifies the picture by eliminating price. It eases estimation because much more is known, or can be guessed, about trends in economic growth and energy growth than is known or can be guessed about the response of supply or demand, and the speed of that response, to large price changes. Furthermore, because the response of demand or supply depends not only on current prices but expected future prices, the simultaneous forecasting of prices and demand depends on methodology that is at best hard to comprehend and usually not available.

But concentrating on gaps and cross-over points is misleading, even prejudicial. It is misleading if it suggests that a gap will actually occur. It furthermore neglects to tell the decision maker the answer to the most important question, which is not the size of some hypothetical gap but how much prices will rise and when. Investors in fuel-economizing equipment, investors in oil and gas pipelines, investors in exploration for new oil and gas, investors in coal gasification or liquefaction technology, household investors in furnace equipment, insulation and solar heating panels, do not know the one thing they need to know when they are given the hypothetical constant-price gap rather than the price. Not only is the price going to be real while the gap is hypothetical; but it is the price, not the gap, that tells an investor whether it is economical to stockpile fuel in advance, to invest in equipment to conserve fuel, or to develop new fuel supplies.

There is a worse effect. By focusing on the gap rather than on the price—by not estimating the corresponding price—there is a strong implication, even if unintended, that price increases are unmentionable. If a variable as obvious and important as price is left out, the reason may be inferred to be that prices would not be allowed to rise, or should not be, and that policy should be based on eliminating the gap rather than anticipating higher costs. A presumption is communicated that a gap should be rationed away, not that a price should be adapted to.

The analysis also inhibits anticipatory action. It suggests that demand will "overtake" supply at some moment in time; and if the conclusion is drawn that prices would *then* rise sharply unless supplies were allocated or demand rationed, there is no scope for taking the anticipatory action that might advance the date of price increases but would moderate their extent.

Finally, the consequences for production and employment are misrepresented if prices are left out of the picture and the metaphor of the "gap" is relied on. A 10 percent gap suggests that 10 percent of the taps will run dry, 10 percent of the engines will stop, and a crucial input to industry, farming and transport will simply not be available when it is needed. If instead it is proposed that fuel is going to cost more in ten years, adaptation can begin early. Nothing special then happens on that hypothetical date when, had prices been held unchanged, the crossover would have occurred and the "gap" would have begun its appearance. Most of the resiliency in an economy comes from market adjustment to changing prices. If you eliminate the changing prices from the picture, you eliminate the image of an economy in which substitution and adaptation are characteristic.

THE ROLE OF GOVERNMENT

Even if the objections to a freer market in energy are satisfactorily overcome, there is still need for energy policy, because there are many objectives that market incentives cannot accomplish. Four deserve emphasis. They are imports, environment, research and development, and the distribution of income.

IMPORTS

There are important costs of imported oil that are not borne by the importer or by the ultimate consumer of imported oil. These are: the balance of payments; the influence of the volume of imports on the world price of oil; the vulnerability to supply disruption; and the need to cooperate with other consuming countries, through the International Energy Agency and otherwise.

Balance of payments. The United States is currently spending almost a billion dollars a week on oil imports. If equivalent funds were spent by oil exporting countries (or by other countries in which the oil exporting countries spent the funds) on currently produced goods and services from the United States, the effect would not greatly differ from spending the same $40 billion a year on an equivalent quantity of domestic fuels: having counted the cost of imports as the cost of energy, there would be nothing further to concern us in the fact that $40 billion worth of miscellaneous goods and services, produced by American labor and capital, was transformed into petroleum by the operation of world markets. But the extraordinary proceeds of oil sales by small countries, recently poor and underdeveloped, has not been matched by an ability to spend such sums in an orderly and effective way on internationally traded goods and services. Huge balances of liquid and semi-liquid assets have grown under the centralized control of a few enormously wealthy governments. Those balances may continue to grow, although not to the startling extent of the 1974-1976 period.

It is beyond this study to analyze the problem for banking and capital markets that could arise from continued increase in centrally held bank deposits and short-term government debt. There is concern, but it is not a concern that particularly affects the consumer of motor fuel, heating oil, or oil-fired electricity. The concern is not reflected in the prices that new domestic fuel supplies would bring on the market, even if those prices were released from control. The market response to concern about these accumulating liquid balances would be through the exchange value of the dollar, interest rates in this country, and the liquidity of the banking system.

Volume and price. Even a modest difference in the amount of oil imported by the United States can make a difference to the price. As pointed out earlier, the difference between eight and twelve million barrels per day—a difference equivalent to about one-tenth of internationally traded oil—could affect the price by an amount that, though not capable of reliable estimation, could be a few dollars a barrel in the 1980s. If instead of eight million barrels at $15 we imported twelve million at $18, the difference would be $24 per barrel on the four million barrel difference.

The higher figure—and this $24 cost is only to illustrate the nature of the calculation—is the cost to consumers collectively of the incremental oil. The importer calculates the cost at the market price (in our illustration, the $18 figure). In recognizing what it is worth to American consumers to avoid importing the extra oil, the correct cost figure is not the higher price per barrel but the still higher incremental cost that reflects the uncertain price increase on the amounts already being imported.

Vulnerability to interruption. If imported oil were a specialized commodity and consumers of it could not readily substitute domestic fuel, importers and consumers could at least be aware that they would be the targets, intended or not, of embargo, sabotage, or war in the Middle East. But if supply is interrupted, the impact, though partly regional, will be essentially national. Particular refineries may be hard-hit, but all consumers of petroleum products will suffer the attendant disruptions. No user of imported products has any reason to believe that his own vulnerability to disruption would be different if he eschewed the imported commodity in favor of domestic fuel. To the extent, therefore, that there is a risk attached to imported oil, because of its potential unreliability on short notice, and especially to the extent that the vulnerability is greater the larger the volume of oil imports, this is a special and additional costliness of imported oil that will not be measured by its market price.

There are, however, at least two ways to insure against that vulnerability. One is to treat imports as more costly, with an attendant reduction in imported quantities. The other is to carry a strategic reserve of quickly available fuel at government expense. As mentioned earlier, the announced plan is to build a federal stockpile of at least a billion barrels by 1985.

Cooperation. There are powerful reasons for cooperating with other importing countries in discouraging the growth of oil imports. One is the corollary of the price calculation given above. If the four million barrel increase in the hypothetical example were not American imports but a combined increase of which the U.S. share was two million, the expected effect on the price might be the same. To continue the illustration, we would be importing ten million barrels at $18 rather than eight million barrels at $15, the difference being $60 million a day. If we could forgo the two million barrels on condition other consuming countries likewise curtail the other two million, we would save $60 million on two million barrels, $30 per barrel.

It is strongly in our interest to utilize our willingness to restrain imports as leverage on other countries, for reasons directly related to the dollar cost of our imports.

Another reason for cooperation centers on the role of the United States in world energy matters. As the largest oil importing country, the richest country, and of all the western developed countries the one richest in energy sources, the United States is considered responsible for leadership in managing world energy problems, world payments problems, and the economic and military security of the western world. The governing board of the International Energy Agency formulated a set of principles that included agreement to make import reduction a central goal of national energy programs. The spirit of any such agreement is likely to have more effect on other countries, the more apparent is the U.S. willingness and ability to demonstrate its own participation.

All these considerations argue for going beyond the achievement of a freer market in fuels toward additional steps to discourage consumption of oil in general and consumption of imported oil in particular. There are, however, opposing considerations that have to be weighed, and no easy conclusion emerges.

- The "side effects" mentioned at the end of the last chapter make it impossible to do as we please about oil imports in isolation from the most vexing issues of domestic regulation and price policy, or even insulated from the macroeconomics of inflation. There is no escaping the fact that any kind of policy aimed directly at oil imports means either domestic price increases, domestic price controls, domestic fuel taxes, or some combination.

- Any action directed against oil imports that is more than nominal in its impact would have to be publicly justified and preferably used in negotiation with other consuming countries in a manner that could be construed as aggressively hostile to the oil exporting countries. The diplomatic dimension might be crucial. The supply response in some of the Persian Gulf countries, and the associated price response, might not be of the kind predicated in the foregoing analysis if the program of import restriction is interpreted as a challenge to OPEC. Resistance in this country to any measures that deliberately raise the price of fuels might have to be overcome by fairly vigorous statements, even overstatements, of the importance of "softening" the world oil market to keep prices from rising even further. The atmosphere would not be conducive to the most subtle diplomacy.

Two other political considerations must be mentioned. One is that the history of oil-import restrictions in the United States is full of evidence that, like so many protectionist and necessarily discriminatory controls, oil-import restrictions invariably carry a heavy load of politics and are at best divisive, at worst severely distorted from any rational national purpose. Even if it were evident that an ideally orchestrated oil-import policy could substantially reduce imports to the great economic benefit of the entire nation, prudence might suggest abstaining from the attempt, or, at least, not assuming that the political process would allow oil-import management to follow an optimum course.

Parallel with that is the likelihood that a rational program to inhibit imports would utilize import duties; sound economic principles suggest duties rather than a mixture of quantitative import licensing, price controls, "entitlements" and other direct rationing techniques. But even a 20 percent ad valorem tax would yield $10 billion per year of revenue if levied on imports alone, twice that much if levied on domestic producers, and still more if domestic gas were treated correspondingly. Ordinarily the fact that a tax, justified on other grounds, also yields revenue would be a welcome side effect; but the recent history of proposed wellhead and gasoline taxes is a reminder that revenues of this magnitude will not be ignored while people choose tax rates guided solely by criteria of sound energy management. What began as oil-import policy may end up as revenue policy.

Thus the considerations both for and against special measures to restrict or discourage imports are powerful. The issues at stake are large. The fact that there are strong arguments both ways does not mean that they cancel out. And the considerations are so diverse in their politics, economics, and diplomacy, that it is hard to reduce them to a common measure. Adding them up and striking a balance would go beyond the purpose of this statement; framing the issues, not resolving them, is the purpose here.

Nevertheless there may be a legitimate way out of what otherwise looks like an impasse. It is to remind ourselves that we are not yet in a position to debate how much further to go, if domestic oil and gas prices have been allowed to meet world market prices, because they have not been allowed to yet.

The first step is to discontinue subsidizing oil imports through domestic price regulation. Most of the more difficult issues arise only after we have eliminated policies that tilt in the wrong direction. Whether we then wish to pursue policies that tilt in the "right direction," taking into account the costs and dangers that would accompany those steps, is a decision that might legitimately be postponed until it is next on the agenda.

It makes sense to watch what happens to imports once the present bias in the price system has been eliminated. We know the direction of the effect; it is hard to estimate the magnitude. Phasing out the present system of import subsidies will begin to provide better evidence of what can be expected, how much more may be needed, and when it is time to consider tilting our price policy in the other direction.

In any event, the possible effect on price inflation would suggest a time-phased program.

A point made earlier can be repeated here. An estimate that the "true" incremental cost of imported oil is higher than the world market price may be an insufficient reason for proceeding to raise the price of oil above that market price; but because it is insufficient does not mean that it is incorrect. It may still be the right standard to keep in mind in judging measures that work in other ways to reduce imports. (There is an important asymmetry to keep in mind about inflation: taxing a commodity to discourage its use raises the price *index* and can trigger cost-of-living adjustments under wage contracts and statutory provisions; subsidizing alternatives to achieve the same purpose, does not.)

ENVIRONMENT

It is almost a matter of definition that the price system

does not reflect "environmental" costs, and they cannot be left to the market. If someone damages his own land without affecting drainage, silting, or erosion on others' property, kills wildlife only in his own pond, runs noisy equipment that no one else can hear or contaminates only his own water supply, he is not said to create an "environmental problem."

The problem is said to be "environmental" when the lead and the sulphur drift downwind to make somebody else sick, the oil spill washes anonymously onto a public beach, the acid drainage from an abandoned mine destroys marine life, or the burning of fuels or clearing of forests change regional or global climate. Environmental effects are the consequences that are *outside* the purview, the cost accounting, the concern or the effective responsibility of identifiable producers and consumers. They are outside the pricing system (except when damage suits are a feasible way to make them costly).

The effects on health of different fuels, or of burning them in different ways, are still little known and the market will not discover them. As they become known, those effects will show up in the market only if regulatory measures are deliberately chosen that make them show up as costs. They may show up as clean-up costs at the point of combustion, as costs of locating where damage will be less, or as the costs of cleaner mixes of fuels and combustion technologies. And they will get costed only when government authority or the legal system obliges the damages to be abated or otherwise taken into account, and only then if a way can be found to assess the relevant costs as guides to action.

There are two problems here, both relating to the way a price system works. One is keeping environmental concerns from being neglected in the marketplace. The second is to keep environmental protection from itself being as divorced from prices and costs as, in their absence, the environmental damages would be.

The environmental costs of energy are large. They will get larger. Most of them have received serious attention only in the last decade, some only in the last few years. There are chemical, epidemiological, meteorological and ecological uncertainties. The uncertainties are not going to be resolved quickly. And some of the problems, like nuclear wastes and endangered species, not only are controversial but invite crusades.

The effect of coal combustion on human health, to take an example, is little understood. Professional opinion about it is undergoing rapid change. Whether the harm is caused by sulphur dioxide in the vicinity of the plant or by photochemically produced sulphates a thousand miles downwind is a question addressed only in the last few years. Sulphur is harmful, but how harmful may not be known to within an order of magnitude for years. Eliminating sulphur from smokestacks adds to the cost of electricity, and with current technology it produces a sludge that is an environmental problem itself. Low sulphur coal is obtainable from the western plains of Montana, Wyoming and Colorado; mining that coal entails land reclamation, scarce water, long-distance rights-of-way, and sometimes the social conflict associated with boom towns.

These are real problems. Most of the reasons why coal production cannot be indefinitely expanded at today's costs relate to environmental and other public concerns, whether it be land use, water drainage, overland transport, millions of tons of sulphur in the atmosphere, or lead and other toxic substances whose effect on health has not yet been studied. Changes in the atmosphere, the temperatures of rivers, the chemical composition of rainfall, and ultimately climate itself are involved.

These genuine environmental concerns are large and important. They will account for a large fraction of the rise in the cost of fuels as well as in the cost of burning fuels. Some of the costs are peculiar to fuel itself—acid drainage from abandoned coal mines or acid rain from sulphur emissions. And some, like power-plant siting and rights-of-way for power transmission, reflect the increasing difficulties of land use in an urban economy with a growing population.

Precisely because these effects are real and substantial, it is important to manage them with attention to the costs of environmental protection itself. Environmental protection is often treated, officially as well as popularly, as an absolute—not as an economic choice, not as a correction applied to the price system, not even as part of the cost of our energy, but as a matter of regulatory standards and prohibitions to be judged and administered without compromise, sometimes as a kind of militant opposition to economic improvement and growth. For some purposes, especially some toxic substances, a purely regulatory approach makes sense. But for most activities relating to extraction or combustion of fuel, environmental damages have to be recognized as costs—costs of abatement to be borne if the abatement is worth the cost, or costs of damage to be borne if they are not worth abating. "Best available technology" is often the standard applied, and it is inherently a standard that cannot reflect costs and benefits and cannot reflect compromise.

People do need protection against lung and heart damage, especially the elderly poor who are most susceptible to whatever the atmosphere brings them and least able to escape it. But the elderly poor also need to be protected against winter cold, summer heat, unlighted stairways, and higher costs of living. In determining the sulphur-removal equipment that power plants must install and maintain, the cost of which must eventually be paid by consumers of electricity, we are determining how much of their budget consumers want to pay for a cleaner outdoor atmosphere compared with heat, light, and air conditioning. Saying that does not settle the issue; it only formulates it.

It is extremely difficult to estimate what those genuine environmenal costs—the costs that will have to be paid or that are worth paying—of fuel and electricity will be during the next few decades. It is even more difficult to estimate how much those costs may be aggravated by failure to treat environmental protection as a legitimate economic problem and to treat it instead as a technological absolute. It is difficult to estimate the costs and delays that may accrue to obstructionist tactics, whether they are legal tactics in the courts or acts of trespass and intimidation. It seems fair to guess that misconceiving the nature of environmental problems, mismanaging the regulatory process, failing to recognize that objectives have to be compared with costs and that environmental values compete with other values, could double or more than double the environmental costs associated with energy. Policy errors of that magnitude should not be accepted as inevitable.

The environmental part of energy is divided among several federal agencies, fifty state governments, and private action in the courts and elsewhere. The issue is not governmental intervention on behalf of the environment so much as it is the mode of intervention, the philosophy of costs and benefits, and the locus of decision.

RESEARCH AND DEVELOPMENT

Research and development, especially in new technologies that are not easily susceptible to patent protection and other proprietary capture by those who invest in the development, are generally recognized as a legitimate concern of federal policy. When discoveries can be adequately protected by patent, copyright, secrecy, or quick exploitation ahead of the competition, market principles are likely to do a good job of inducing the economically justifiable research and development to take place. But when the discoveries and the experience cannot be capitalized by the investors—when the investment generates mainly a public demonstration of feasibility (or infeasibility!), when the development is a "learning process" that people can carry away with them, or when part of the learning is the discovery of environmental concerns that, once identified, are visible to all—the results of research and development will be undervalued in the market.

This principle can apply to a broad range of initiating activity, from basic research at one end of the spectrum to exploratory development, testing, prototypes and pilot operations, demonstration plants, even pioneer operations on a commercial scale. But like the arguments for immediate restriction of imports, this argument for federal subsidy of new technology deserves a guarded response. It is easy to exaggerate the need for governmental sharing in the cost of a new exploratory production process. Compared with most technological development done at government expense, e.g., military and space technology where the government itself is the consumer, development for consumer markets is an open-ended affair.

Nevertheless, in energy technology, especially new technology for liquid fuel and gas, there are special reasons why exploration and development, even initial experiments with commercial-scale production, can have a national economic significance beyond the criterion of profitability.

One is the importance of arriving at a reduced range of uncertainty about the nature of the energy problem itself. Just knowing whether or not some important synthetic fuels would eventually be competitive, or knowing the world oil prices at which they would become competitive, could help to avoid serious mistakes in energy planning, both private and public. Private investors only lose by investing in a plant that produces mainly the valuable information that such plants are not yet competitive and are not going to be for many years.

The same principle applies to exploration. From the point of view of a private firm exploring for new fossil fuel deposits, success consists in *finding* the deposits that exist. For formulating national energy policy, exploration often has a value in *finding out*, whether the findings are positive or not. The government's National Uranium Resource Evaluation Program (NURE) is based on this principle: it is an attempt not so much to find any uranium that exists but to get a better global estimate of how much uranium there is to be found, at different concentrations and extraction costs. In the same way there is a national interest in knowing how much natural gas and petroleum is going to be found, not only in United States territory but worldwide, because so many decisions, public and private, depend crucially on overall estimates of the likely quantities that may become available, at different extraction and distribution costs, over the coming decades. Knowing only that there is an abundance of gas, or alternatively not much, to be found, is of some help to the company that explores for gas but not nearly as much help as knowing where it is to be found. But the same information is especially helpful to investors in, say, coal gasification, just as it is helpful in deciding on an oil import policy. These decisions depend mainly on knowing what is going to be found, not where to go look for it.

In the same way, learning what the production and environmental costs of coal-based gases and liquid fuels will be, or oil from shale, can provide a crucial parameter that helps to put boundaries on the nature and magnitude of the energy problem which will face this country in the 1990s and the early years of the next century.

There are special reasons in energy why the development of a better knowledge base is of national interest. It is often as important to know that a particular technology will be environmentally unacceptable as to know that it will be acceptable, or as useful to know that a technology will not help in holding down the price of liquid fuel as to know that it will help. Private investors get no return from negative results; but bad news can be a valuable warning to others.

A powerful argument for a strong government interest in the development of new technologies for liquid fuel arises from the markets' undercosting of imports, as described earlier. It was argued above that the savings due to reduced imports could substantially exceed the nominal price per barrel. It was remarked that while that provides a strong argument in favor of import controls, there may be powerful countervailing arguments. But whatever policy one elects with respect to import controls, the higher cost is the correct one to have in mind in considering alternative energy decisions. Synthetic liquid fuels, for example, at a premium above the world price for oil, could be worth their cost if they reduce oil imports, even though consumers would not pay that price because the consumer who pays the full price gets only the nominal value, while the rest of the value accrues to the entire economy in lower oil imports and a possibly lower price of oil.

This could, of course, be a general argument for subsidized domestic production of liquid fuel. While permanent large-scale subsidization of commercial production could be objected to on a number of grounds, at least the federal cost-sharing or subsidization of the relevant research and development could properly be considered justified by the excess of the true collective cost of liquid fuels over the nominal world price of oil.

INCOME PROTECTION

When prices change incomes change. They change because earnings are affected by prices; and more generally they change because price changes have different effects on what different people can buy with their incomes. An increase in the price of coal can reflect greater earnings for coal miners or greater earnings for companies that own mining properties; it also reduces the real income of people whose electricity costs more. If all fuels become more expensive most of us—not all of us—suffer reduced real incomes, but not all in the same proportion; some of us are old and need warmer homes; some drive longer distances to work or have more children's clothes to wash. The same is true for all prices. But when particular clusters of prices, like meat or medicine or fuel or house rents go up or down, and especially when they go up, there are identifiable effects on people of different incomes, ages, and locations. When the changes are substantial, as with fuel, there is an expected tendency for the people who are most disadvantaged to try to protect themselves through government intervention to forestall the price increases. And because fuels are comparatively standard commodities, already in most cases subject to taxation or to regulation with respect to interstate transport, and because fuel and electricity prices are directly visible to consumers, income protection becomes a politically powerful argument for price control. (Witness the periodic popularity of gasoline rationing whenever "shortages" appear or seem imminent, i.e., whenever prices appear about to go up.)

The price system, when it works well, is impersonal and indiscriminate. What it does not do, and what should never be claimed for it, is to bring about the distribution of income that we might prefer on general grounds of equity and social welfare. The price system that determines our individual wages, salaries and profits, and our individual costs of living is attuned mainly to the supply and demand for particular goods and services, and it generates the distribution of income as a hugely important byproduct. (People with greater needs can sometimes work overtime; people can raise their incomes by moving to where their own particular talents would be in greater demand; people can raise their future income by saving in response to interest rates and investment opportunities; but in determining the relative earning abilities of thirty-year olds, fifty-year olds, and seventy-year olds, the market does it the way it determines the relative prices of avocados and oranges —through supply and demand, not considerations of social welfare.)

There are, broadly, two altogether different kinds of mechanisms for changing the distribution of income or for protecting the existing distribution against change. They can be called the microeconomic and the macroeconomic. The microeconomic mechanisms change income distribution by intervening in particular markets, holding prices up (agricultural price supports, minimum wage laws) and holding prices down (natural gas, rent control) and protecting markets from competition (tariffs and non-tariff trade barriers, airline regulation and taxi medallions) and sometimes by regulating markets or providing information services to make markets work more competitively. The macroeconomic measures work on incomes directly, and in the aggregate rather than with respect to particular markets; these are income taxes, social security, welfare, and sometimes benefit programs for particular groups like veterans, the blind, or college students.

As a general approach to income transfer or income protection, there are two powerful reasons for favoring the macroeconomic approach. One is that it does not so much distort the price signals and price incentives that coordinate demand and supply. The second is that it is much more likely to protect incomes or transfer incomes in accordance with principles that might command widespread political assent. Macroeconomic programs can target the poor, the elderly poor, the very poor, the disadvantaged, the disadvantaged poor, and any other groups within the population who can be defined by reference to what makes them particularly needy or what gives them special claim. Only occasionally does a microeconomic intervention provide help to a substantial part of the target population and only to the target population: some things may be purchased only by, say, the rural elderly poor, and subsidized provision or even price control will concentrate the benefits on the target population if they are indeed the target population. Holding down the price of gasoline or heating fuel, especially if it means that some people cannot even get connected to the cheap natural gas that would be a bargain at twice the price, distributes its purported benefits over a large segment of the population in a way that is hard to calculate and that is unlikely to correspond to any acceptable criteria for income protection and income support. The fact that the poor buy gasoline too—not all of them, especially not the very poor and the especially disadvantaged poor—does not make price controls on petroleum a program to help the poor. The poor who don't drive are likely to be as much in need of financial help as the poor who drive, and gasoline price control neither confers most of its benefits on the poor nor confers significant benefits on most of the poor.

There are special cases in which sudden price changes would have mischievous effects on particular groups that have a special claim to protection. There are cases in which the government, wisely or unwisely, made price commitments that perhaps should not be abandoned abruptly. (The control of inflation, as mentioned earlier, may occasionally demand direct action on particular prices.) The microeconomic approach to income protection therefore can often be justified as a special case. But compared with the more explicit and more effective macroeconomic route, managing and manipulating the price system to preserve or to affect the distribution of income ought to be justified in terms of special cases. Primary fuels, refinery products, and electricity have a price incidence that is spread so broadly over the population at all income levels and in all regions and at all ages, that for purposes of income protection the justified special cases should be rare.

THE ONE CERTAINTY: UNCERTAINTY

A special principle underlies any approach to energy policy—*the principle of uncertainty*. No one really knows how much undiscovered fuel there is, how quickly it will be discovered, how much it will cost to produce and what the environmental effects of consumption will be. It is not likely that uncertainties about resources and the costs of using them will be dispelled within a decade, or even two. We must design policies that admit the possibility of surprise and that weigh the relative risks of being caught sometime in the future with unanticipated good news, or unanticipated bad news.

We have similarly poor information about how the economy responds to changes in the prices of energy. Since the end of World War II there have not been alternate periods of markedly contrasting energy prices to give us "experimental data" about conservation, substitution, and the stimulus to invention and exploration. There are technological uncertainties about the cost and safety of burning different fuels for different purposes in years to come. There are grave uncertainties about the security of oil from the Persian Gulf, the Mediterranean, Southeast Asia and Latin America.

Given all the uncertainties, the wisest course is to pay special attention not only to the policies but to the policy *process*. Long-term targets may be needed for planning, but they must be susceptible to short-term revision. Buffer stocks of petroleum, for example, can allow not only a cushion against sudden shortages but an even more important cushion against decisions made in haste.

With full recognition that the market cannot respond to all of the environmental or foreign policy considerations, it remains true that for most business and consumer decisions the market, as a process, has the important virtues of flexibility and adaptability.

ABOUT THE AUTHOR

THOMAS C. SCHELLING, professor of economics at Harvard University since 1958, has been a faculty member of the Center for International Affairs and of the John Fitzgerald Kennedy School of Government, chairman of the School's Public Policy Program, and, since 1974, Lucius N. Littauer Professor of Political Economy.

Prior to 1958, Professor Schelling was an economist with the U.S. Government in Copenhagen, Paris, and Washington, D.C., in foreign-aid programming, and professor of economics at Yale University. On leave from Harvard, he was with the Rand Corporation, and later with the Institute for Strategic Studies in London. In 1976 he was a Lady Davis visiting professor at the Hebrew University of Jerusalem.

Professor Schelling has been a consultant to the Departments of State and Defense and to the Arms Control and Disarmament Agency, and a frequent lecturer at the Foreign Service Institute and the several war colleges. He was project director for the CED national security subcommittee which issued three policy statements: *Military Manpower and National Security* (1972), *Congressional Decision Making for National Security* (1974), and *Nuclear Energy and National Security* (1976).

He is the author of *Micromotives and Macrobehavior*, 1978; *Arms and Influence*, 1966; *Strategy and Arms Control* (with Morton H. Halperin), 1961; *The Strategy of Conflict*, 1960; *International Economics*, 1958; and *National Income Behavior*, 1951.

He received his B.A. from the University of California, Berkeley, and Ph.D. (Economics) from Harvard University. In 1977 he received the Frank E. Seidman Distinguished Award in Political Economy.